Contributing to Healing

daniel perret

Nature healing, distant healing, spiritual healing, prayer, collective action

© 2020 Daniel Perret
Edition: BoD - Books on Demand GmbH
12/14 rond-point des Champs Elysées
75008 Paris, France
Druck: Books on Demand GmbH
Norderstedt, Deutschland

Dépôt légal Août 2020
ISBN 9782322237562

Cover photo
Celtic circle neat Lough Gur
Arc Images, Cathal Ryan
Co. Limerick, IRL

In this book I am focusing on how we all can contribute to positive change and relief of suffering by directing positive thoughts and wishes towards others. We can do this whoever and wherever we are. This compassionate energy movement out from our hearts is not only beneficial for others but also for ourselves. I wish to share my experience by giving an overview of various forms of distant healing and prayer.

There is an increasing debate about the rights of nature all over the world. There is also a growing awareness and urgency about the numerous crises that need to be resolved in the near future.

This book came about with the arrival in Spring 2020 of a new kind of nature spirits, the Lenos beings, They are directed by the Celtic Goddess Bríd, a daughter of the Tuatha Dé Danann, the people of the Goddess Danu that lived in Ireland before the Gaels. The Celtic Circle on the cover is located near Lough Gur, Co. Limerick in Ireland and was likely to have been a ceremonial ground for solstice and equinox celebrations. The circle symbolizes, as with the North American Indians and many other native people, the wholeness, a community of equal beings and the unity with the All-One. The main task of the Lenos beings seems to be to encourage human beings to send daily distant healing to nature in the surrounding area.

My thanks go to Marie for editing and discussing.

Contents

1. Getting started

2. The practices

3. Reflections

Introduction

How can we, in these times of great change, use our thoughts and actions for the wellbeing of nature and of all sentient beings: animals, humans, bees, birds, nature spirits…

I have been studying and teaching spiritual healing for many years. The changes in the world happening in recent years have very naturally shifted my focus from personal development towards nature and earth healing. My website and my book 'Earth Healing' (in French and German) describe this transition.

Healing in all its aspects, from spiritual healing, hands on healing, distant healing, and prayer, are all straight forward and natural procedures. They are essentially based on the universal energy of love. In all these years of practicing various forms of distant healing I have been able to observe, that despite this simplicity, we can meet a number of inner hindrances in our endeavours to contribute to healing. Good intentions are not enough. The spiritual and therapeutic discoveries of the recent decades have brought us tools that allow us explore and transform such hindrances in depth.

All healing treatments work through realignment with the All-One because we are all part of this. Our time and the times to come will generate a lot of compassion and mutual help. We will also have to focus much more on nature and all its visible and invisible beings, in order to heal the disharmonies and damages human beings have created out of greed, ignorance and lack of compassion. In this process we are not alone. Nature is patient and good willed. Many nature spirits and beings of the divine field are waiting with outstretched hands willing to help us.

In recent years I have come across new kinds of nature spirits. Their task consists of furthering the co-operation between humans and nature spirits. The Elementals of the 5th kind appeared in 1993 and other new nature spirits called Lenos-beings appeared this spring 2020. Their task is to facilitate distance healing between humans and all aspects of nature (animals, insects, plants, nature spirits of all kinds). See more page 54

My purpose in writing this book is to underline the necessity and simplicity of using distance healing and especially bringing the focus to our own neighbourhood, taking responsibility for the natural surroundings beyond our own private garden. This is the area upon which we can have a direct influence. Taking care of our private garden and private life is no longer sufficient, if we want to save our planet, our climate, our animals and nature in general. We need to join up with others and take collective action where needed for the quality of our lives.

Behind every manifestation, visible or invisible, there is a consciousness and thus a conscious, feeling being. We must learn to acknowledge the work of invisible beings, thank them and include them in our decisions about nature. We cannot work for nature without talking to its beings, as this would be a condescending, paternalistic attitude. It would be not far from wanting to help children in slums or native tribes without ever talking to them.

A great many health practitioners of all kinds are specialised in caring for our health and providing healing. Their work is indispensable and is often directed at alleviating physical symptoms. In the following pages I will not address this kind of physical healing. My field of competence concerns spiritual healing and the work with subtle energies.

The difficulty with spiritual healing is that it often does not produce immediate visible effects on the physical body. Despite the fact that healing is working through high frequency energies of the divine field, the hindrance to healing is due to own limitations. The results of spiritual healing are mostly found on a subtle level and are not always easy to observe. How then do we know, when we send healing or pray, whether our contribution to a healing process has been useful?

In my research on subtle energy I have observed three types of effects. In all three cases people present could confirm this. In well over a hundred cases of distant healing concerning nature spirits of all kinds, I could see how the energy line coming to my crystal from a specific nature spirit disappeared after I used my thoughts and feelings in distant healing. In addition to that perception I could get confirmation from the nature spirit, that my intervention had produced the effect they had wished. (see pages 44)

Whilst sending distant healing with my harp I perceived the second type of effect. After the distant healing I could observe a wave of circular energy spreading out starting from our house and spreading very far out on Google maps. (see pages 74)

These two examples show the effects of how our wishes and thoughts operate in prayer and distant healing. They also affect us and our health. A lot has been written about self-healing, including in some of my books. I won't go more into this here.

I could observe the third kind of effect in my own body and energy field when spirit beings were performing energy operations on me. I felt significantly better after the operations and could feel that they had created deep and positive changes, even breakthroughs within me. I could sometimes also feel rapid energy movements in the life ether near my body.
(see illustration on the human etheric page 105)

My sources are my studies with the spiritual healer Bob Moore in the eighties and nineties, my personal experience and guidance from C, the college of spirit beings of Rocamadour. They form a think tank that includes experts with deep knowledge and insight into the divine field. I presented the 12 members of their inner circle in my book *Earth Healing*. This present book is the seventh book I am writing with their help. I communicate with them essentially through yes/no questions and my Hartmann-antenna, a kind of pendulum device.

Another source are my interviews with nature spirits like the elemental of the 5th kind near our house.

My thanks go to all of them.

1. Getting started

Our ambivalence towards the spiritual universe

I mention on page 93 an example of a collective thought form that has been hindering our contact to the wisdom of the universe. I also mention our damaging thought structures concerning nature and animals. In Europe the history and ambivalence towards the spiritual has left us with a conflicting relationship to the church. Collectively we have confused the 2000-year old history of Christianity and the church with the spiritual and the Universal Consciousness. One is a human organization with its history and tragic mistakes, the other is the source of our unity with ourselves, nature, the earth and the Divine. Healing of this deep, painful division in us with the Universal Consciousness is necessary on a collective and individual level.

Co-operation between the Worlds

For a number of years, I have been observing changes in the energy on earth. I have written about the elementals of the 5th kind and the Dagdas in my book *Earth Healing* [1] and also in English on my website. The energy changes are often not induced by human behavior but by beings of the divine field. I have drawn much inspiration from the forty books published by the *Flensburger Hefte* with intelligent and stunningly normal interviews with nature spirits of all kinds. These interviews have laid the foundation of an enlarged science describing the co-operation between nature, nature spirits and humans.

Individual actions

For the sake of completeness, I will mention the obvious individual actions that we need to do such as picking up garbage in nature, refraining from buying things that have negative effects on the environment, taking care of our gardens and plants at home, and the awareness of the consequences of our consumption of certain goods.

Collective actions

In many places around the world groups of citizens are working to obtain legal rights for river systems, mountains, protected forests, and nature parks. This is a form of distant healing as their efforts go towards an ecosystem and its beings. In recent years several river systems have got a legal status of a non-human individual: Whanganui River (AUS), Ganges (India), Klamath River (USA), Lake Erie (USA).

In Europe several citizens groups and NGO's are working for the rights of river systems: e.g. Rhone, Loire. *The Global Alliance for the Rights of Nature GARN* and its European branch the *European Hub for the Rights of Nature* are examples of such NGO's. Collective cleansing actions for riverbeds, beaches as well as the planting of thousands of trees are other examples of a growing awareness towards the direct surroundings.

This movement has led to changes to the constitution of some countries. Furthers steps are needed to translate that into laws, court sentences and concrete actions so that the healing of nature to come into effect.

Large water elementals are in charge of whole river systems and are geographically anchored in precise places. They direct a myriad of large and smaller water elementals in charge of confluents, smaller rivers, lakes, and swamps connected to the same large river system. In the following map of France these large elementals are indicated with white stars. The circle in the Golf of Gascoigne shows the area and section of sea cared for by this large water elemental.

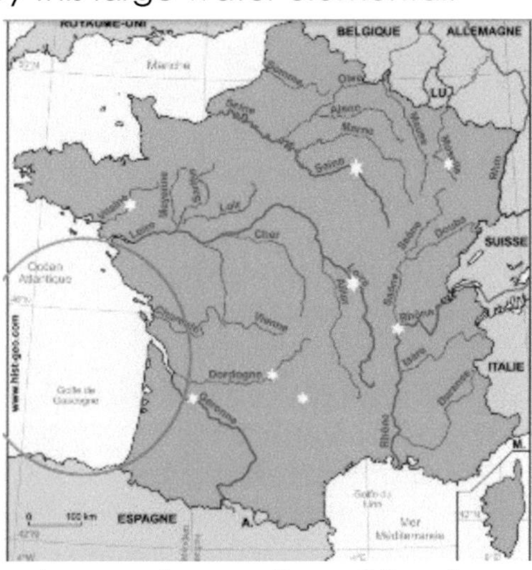

We can communicate with these elementals, as indigenous people all over the world and throughout the ages have always known. All nature spirits are conscious, intelligent, and sentient beings. They have a great knowledge and wisdom concerning all that is happening and has happened in their river system. The large water elemental of Dordogne river told me some months ago about a precise section of the river in the county of Corrèze. He said that he unusually abundant rains during the winter months had brought enormous amounts of earth and mud into the river and could potentially create dangerous blockages in the flow. Communicating with these beings helps us to understand how they work and what the present challenges are. What they tell us is often not immediately visible to our eyes but brings us to the pulse of a living organism.

Thomas Linzey wrote in his article : *Rights of Nature: One Big Step for Nature, One Small Step for Humankind*: "Over the last five years, courts in India, Colombia, and Bangladesh have declared that rivers and other ecosystems have rights, and political parties across the globe – including the Democratic National Committee (DNC) in the United States – have declared their support for rights of nature laws.

Tribal nations have also led the way. The White Earth Nation of Ojibwe has adopted a law recognizing the rights of manoomin (wild rice), and the Menominee, Yurok, Nez Perce, and Ponca Nations have adopted laws and resolutions recognizing the rights of rivers and other ecosystems. As they have expressed, this represents both an issue of sovereignty, as well as an alignment of tribal laws with long-held indigenous respect for nature."

Published July 10, 2020, by Common Dreams

Another type of collective action is shown in the photo below of Fayetteville in North Carolina, June 2020. It particularly moved me. Instead of a violent confrontation between the 'Black Lives Matter' movement and police, the latter knelt down out of respect for the

A group of about 50 police in Fayetteville, North Carolina, showed their support for Civil Rights protesters kneeling in the streets before marchers on Monday. BY FAYETTEVILLE POLICE DEPARTMENT

demonstrator's cause. This completely transformed the usual animosity and confrontational attitude. This collective action in Fayetteville showing respect for racial equality attracted positive energies from all over the world, during a very difficult time for the whole nation.

Thoughts about the Divine Field

I will begin by sharing the definitions and beliefs I will build on in this text: Healing, God, Spirit, Consciousness, Whole/Holy, Free Will, Divine Harmony

Healing

A process that aims to restore the original Divine Harmony, Unity and Wholeness.

The word healing is related to the words holistic and holy, as well as to the ancient Greek term 'holos' (whole, and itself is part of a larger whole). The word hologram is interesting as a concept, as in each part, the whole is always recognizable. Seen like a hologram a human being is a whole and at the same time part of a series of systems wider than him or her. (environment, neighborhood, universe).

Depending on the therapeutic concept Healing, as a restoring of a wholeness, refers to different concepts of wholeness: body-spirit, feeling-thinking, incarnated self-timeless soul, male-female polarity, and the unity with the great Whole, the Divine Field, the Universal Consciousness. Healing is not concerned with physical health alone, but tends toward a physical, psychic, and a wholeness of the timeless spiritual Soul.

Spiritual healing, distant healing and prayer are simple: we (the sender) aim at connecting the Divine or Universal Consciousness within us to the Divine Self of the receiver. The Divine Self is the most natural part in us, connecting us to all, and this aspect is free from any restricting layers of the Ego.

God

As soon as we ask: what is God, what is divine energy, what is the Divine in us, who or what is being healed, what is Spirit, can everybody do that, are other energies sent in distant healing besides the Divine ones? Who has a Divine Self and how can we contact it? Do these forms of healing and prayer exist in all religions? Who do we pray to? we touch on complex questions for which there are no ready answers.

The Old Testament image of a punishing God is outdated. It was the result of unconscious projections of people at that time in history. These negative projections were created by their own conscience due to their lack of understanding and their conflicting behavior with the laws of the universe.

The following sayings come closer to reality: *'The way you make your bed, is how you will sleep'* and *'The way you call into the forest, in the same way the echo will come back to you.* Life functions within laws that operate here on earth. When we want to row across a lake or fly through the air, we need to respect the laws of aerodynamics or hydraulics. In a similar way we need to respect the laws of Truth and Love because no thought, no action ever gets lost in the world's memory bank. Sooner or later we are brought to face the disharmonies we have created and rectify or heal them.

I have chosen to replace the word God by the Divine Field as it was shown to me by 'C', the college of spirit beings. This divine field is filled with a great number of spirit beings. [11] See the list of the 21 spiritual spheres and their beings on page 103 at the end.

In the course of time human beings have prayed to very different beings in this field and thought they were praying to God or Gods. In my understanding the spheres 17-21 were often addressed as being 'God' in olden times. These are the Seraphim, the Cherubim, the Throne angels, the Cupidos (high angels of Art, Love and Beauty), as well as the top level 21 where we find for example the Black Madonna, Christ, Mohammed, and Buddha.

The Divine Consciousness and possibly the Black Madonna as the Earth Goddess, may be above that level. I have heard the Black Madonna talking to me, one experience was also visual where I saw her face in front of me. [10] I have also witnessed my friend Eva Høffding's channelings and dialogues with the Black Madonna and with other light beings.

This contact with spirit beings of the divine field has changed my way of seeing prayer. I can direct my requests to a certain spirit being whose field of competence is required. I have for instance asked a specific being to help me improve a physical condition. I then received a number of energy operations and I could sense what was being done.

I direct prayers of a more general nature to the 'Great Spirit' or 'universal consciousness' but sometimes also include spirit beings of the divine field.

See page 78 for the rest of the definitions: Spirit, Consciousness, Whole/Holy, Free Will, Divine Harmony.

My personal beliefs

I believe life is a continuous learning and healing process and this is the purpose of being alive on earth. Our timeless soul has to a large extent chosen before conception, our life program including all the lessons and hurdles, that we have planned to handle and learn from. Our life plan would include the choice of parents, siblings, birth place and the main circumstances and events of our life. This belief has the advantage of giving all the power and responsibility to individuals for changing their lives as it situates the causes with us and not with an external authority.

From my experience of a number of past lives, I could also detect life overlapping themes with people that can take several lifetimes to come to terms with. Learning how to integrate universal wisdom into our individual souls is a complex task. During our lives new themes emerge that all ask for a healing approach. Our life is a lifelong learning and healing process, a ripening of our timeless soul.

A healer can and should only intervene when having some understanding of the personal background of the person requiring assistance. The intervention needs the permission of the timeless soul of the individual and the beings of the divine field.

In order to be aware of our responsibility as a healer we need to be clear about our own motivation. If we try to impress someone else or wish to be superman or superwoman, we enter dangerous waters.

This mandala of the chakra of liberation is also called the life force chakra. Published by Rigdzin Jigme Lingpa as part of the Longchen Nyingthig terma cycles.
This mandala of the Tibetan Buddhist tradition has an extraordinarily high and pure energy. This energy is composed of divine and upper mental energy.

The essence of the divine and universal consciousness

Here a new definition of God:
The essence of divine consciousness, the universal and wise balance is a limitless power, within which the equilibrium-harmony results from a mutual witholding of the universal impulses: of the male and the female, the intelligence and the love, the will to experience and the will to submit, the wish to be active and the will to withold and maintain the momentary state of equilibrium in order to build binding and stability.

The divine consciousness of the universe is the source of love, beauty and creativity. In that sense 'God' is not personified as the Buddhist concept of vacuity.

The same way that light shines upon and reveals our blind spots, these original energies are neither condemning nor punishing.   See Christ letter 7

Prayer and distant healing try in a natural way to re-establish the connection to the universal intelligence. Yet, what role do we see 'God' playing in all of this? Distant healing, meditation, and prayer have as a precondition concentration, quietening the mind 'the inner radio', transformation of our ego and fears.

How to continue with old expressions?

"May God protect you" - Now that we may have revised the concept of a personalized God, we can no longer direct prayers to 'him'. We can address our own responsibility and possibly some beings of the divine field. See the 21 spheres of the divine field and its beings at the end. We could for instance chose the following wording: "May the Divine protect you".

"God bless you" - A blessing conveys the energy of the Universal Consciousness towards a person or a place. I would tend to use "May the Divine bless you".

...the highest contact with energy

What did my teacher Bob Moore mean, when he said?
"The reflector ether is the highest ether, the most advanced contact we can reach with energy." This seems the key to understanding our role in distant healing. Using the word 'reach' means it is the transformation of our ego (lower astral with its painful emotions, and lower mental) that enables us to use the higher astral with its finer feelings and the higher mental with the immense potential of the pineal chakra, intuition, clairvoyance, etc. thus bringing us into contact with higher energies and beings of the divine field.

(see page 71 for the complete quote)
See for more on 'astral aura', 'mental aura' my book [7]. The mental aura is represented by its only stable structure that is the rim of the mental aura. The lower astral aura contains our painful emotions (nearer to the body) and in the upper astral aura our feelings.

The following transformation can be done with any prayer regardless of tradition and religious context.

The Lord's Prayer

Any prayer should not become part of a meaningless ritual. The wording of the Lord's Prayer has come to us from another time and needs to be adapted and brought closer to us. I now include my translation with a non-dualistic approach, that respects the Divine in its new light. As the Divine is not separate from us, we can no longer direct a request to something outside of us. This allows us to take back our own responsibility and not project it onto someone outside and then wait until 'the other' finally does what we asked for.

Our Father that art in Heaven (...art in all)
Hallowed be thy Name
The Universal Consciousness has not got any name. But we can address the Universal Consciousness in a sacred, holistic approach.

Thy kingdom come
This can only be understood as the manifestation on Earth of the values conveyed by the Universal Consciousness.

Thy will be done
Do we at all have any idea, what the will of the Universal Consciousness really is? We can actually only refer to universal laws and values. These are the 'will' of the Universal Consciousness and we need to include that we may realize ourselves and fulfill our soul purpose while respecting our own free will.

Give us this day our daily bread
'Bread' very likely means spiritual nourishment we get daily when we 'let in' the Divine Consciousness as it continuously manifests through creation.

Help us to forgive our own trespasses
As we forgive them that trespass against us
As soon as the erroneous concept of a Divine Authority dissolves, we can easily see, that there is no 'Him' that could forgive us. The most difficult task for us, is to forgive ourselves and see where we have transgressed the laws of the universe.

May we not fall into temptation
And may we free ourselves from evil.
Again, it is not a 'He' that will protect us and lead us into temptation. The previously described mechanism of the ego leads us naturally to test different paths. We are asking therefore to clearly see, before engaging onto a dangerous path outside the divine laws. We can ask beings of the divine field to help us in this.

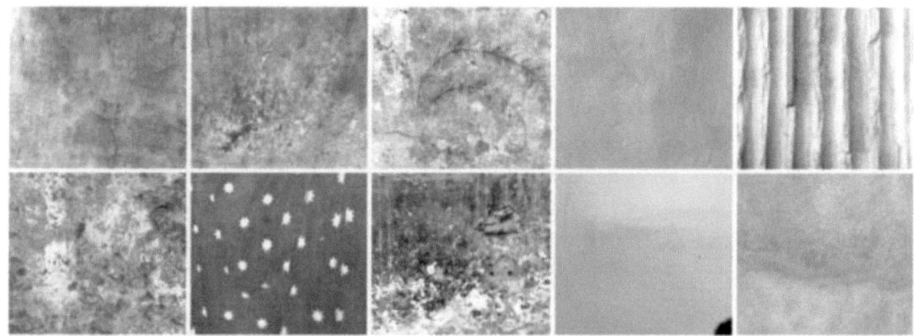

Walls of Rome

2. The practices

Distant Healing

Distant Healing is comparable to prayer in its way of functioning. Scientific studies have proven that the effects of prayer can be measured over long distances.

Distant Healing is different from prayer because the receivers need to ask for it and be open to receive the energy transmitted by the (group of) healers. Respecting the free will of a person is central.

There are though some exceptions to this: some countries, conflict areas, animals, and nature spirits can be included as it is not always possible to ask them for permission. Persons who not able to formulate a request for healing can be added: for example, small children, people in coma, and people recently deceased.

At the beginning of distant healing we silently formulate our intention to send healing energies to persons or nature, followed by a period of meditation during which we do not interfere, letting the divine field taking care of conveying that energy.

Directed distant healing is directed to specific persons or nature spirits, who have asked for help and cannot be present. Individuals or a group of people, e.g. healers can generate distant healing. This can happen at a specific time of the day or in the week.

For many years distant healing has been a regular part of my daily activities. I am part of a group of healers, who send distant healing to the people on a list each morning between 8.00 - 8.15. This list is kept up to date by a spiritual healing Center. I also have my own list of names of people to whom I direct healing. People on both lists have asked to have their names on the lists.

Distant Healing – the procedure

- Centering in our heart
- Contact the individuality point*
- Come back to the heart
- Expand the energy towards the front in a widening way
- Using a feeling contact to the color gold
- Asking for help whoever we want: the universe, high beings of the divine field, Divine Source or Universal Consciousness
- We can then relate to the list of names that we intend to include in distant healing.

*) see S. 92, and my book [7]

"Understanding the pattern that set the stage for her dis-ease made my client whole again. She was healed and ready to leave this earthly plane, peacefully."
Laurence Marie in Harp Therapy Journal, spring issue 2020

The effects of distant healing

For the persons involved distant healing is essentially based on trust and faith. Some may feel the energy generated, whilst other people myself included, do not feel so much. I know that I have a personal way to perceive energy so this does not make me doubt distant healing. Each person perceives energy in a different way and one way is not better than the other.

I have, in cooperation with C, measured the energy transmitted by the healing group on April 15th 2018. I am using the Bovis Scale of measurement. Even if it may not be an objective scale, this method provides us with a good way of comparing several objects or moments in time during distant healing as shown in the following graphics.

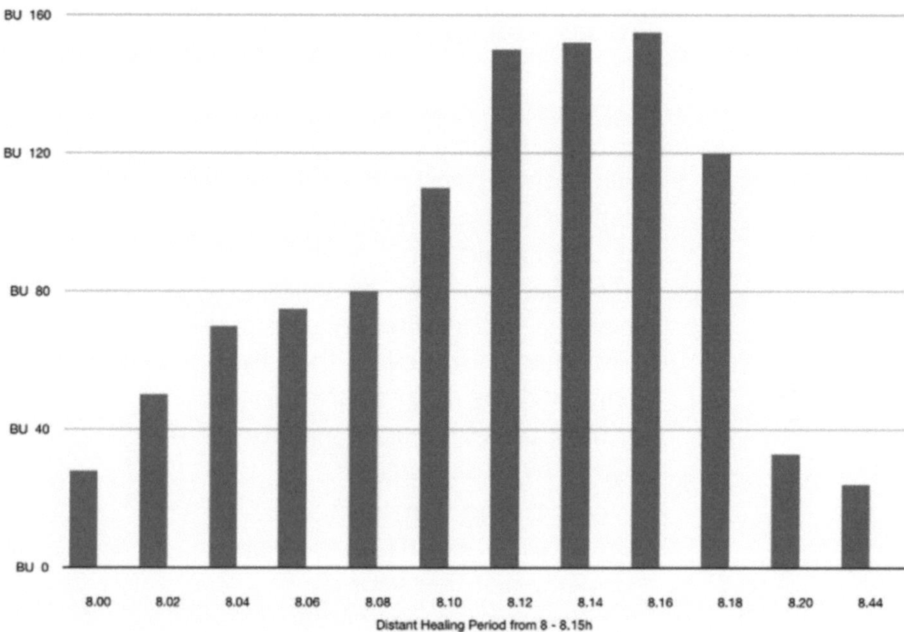

Measurements of the group's distant healing energy in Bovis units, 15.4.2018
The units are kept here in 1000 BU (Bovis units).

In my personal experience measurements above 100' BU are relevant. I have also checked the measurements with my Hartmann-antenna using the vertical distance between my left and my right hand with the antenna. This distance corresponds usually to 120' BU for a meter. According to my personal experience and comparing it with other places and situations, the measurements here are quite high. I can sometimes feel this energy level in a mediation as a type of 'inner electrical' state during which I am completely alert.

Distant healing with the harp [2]
5-10 minutes

In spring 2019 I was shown in great detail, by a group of spirit beings that I call a think tank for harp distant healing, that harpists could send a particularly effective form of distant healing. At that time the nature spirits communicated that they preferred me to play the harp for their distant healing instead of just my usual prayer.

What is being transmitted through sound? The receivers cannot hear the sounds. Melodies and even playing techniques are unimportant. What remains is the depth of feeling in our heart.

The effect of this distant healing spreads over the whole planet and can also be directed to precise beings. Harp and harp like instruments are especially appropriate: hammered dulcimer, zithers, sitar, koto, cembalo, monochords, and suchlike. The essential aspect is the many resonating strings. A guitar or violin would not have the same effect.

Prayer

... directing of our thoughts towards something beyond the limits of our usual consciousness. Opening a window towards a wide, well-meaning Unknown. A request, a prayer to be more deeply embedded in the All-There-Is. Prayer as a silent dialogue with the Divine Dimension, the Infinite, whoever or whatever that may be and that we are an integral part of.

Prayer is connected to our beliefs and our trust that someone may hear our prayer. To whom should we direct our prayer? Are there different beings receiving our prayers?

In distant healing we act as a link, a facilitator, who because of our physical existence on earth, can help to bring divine energies down and pass them on to beings in need. With prayer it is usually not necessary to ask the receiver for their permission.

It may be helpful to distinguish between three types of prayer:

Self-centered prayer can cover a whole range of prayers from a suffering prayer, a request for help for pain, up to requests for personal benefits. This type of prayer is essentially ego motivated.

Altruistic prayer can arise from different motivations, extending from egoistical, hoping for personal benefits, to the genuine altruistic, selfless prayer. With this latter it becomes very similar to distant healing.

Mystical prayer is motivated by a search for a deeper connection to the Divine Field or Universal Consciousness.

Centering prayer is according to Thomas Keating* and Cynthia Bourgeault* a contemporary expression of the contemplative Christian tradition as it was developed by the desert fathers and mothers. This form of prayer was further developed by mystics like John Cassian, the anonymous author of 'The Cloud of Unknowing' or John of the Cross and St. Teresa of Avila, to mention just a few. Each religion knows of comparable forms of prayer.

Contemplative = an interiorized way of experiencing

This type of centering prayer is sometimes called contemplative or receptive prayer. During a first phase it is a method of relaxation, of emptying of one's thoughts, a letting go of worries and criticizing others or ourselves. This allows some stillness and inner calmness to spread. Patience is needed, sometimes repetition of words or images, as well as the practicing of the receptive 'Not-Knowing', during which we endeavor to experience within the presence of the Divine or Universal Consciousness.

Centering Prayer
Or gathering prayer,
20-40 minutes

The aim of contemplative prayers is to harmonize body and mind and allow through an intuitive-feeling attitude the Universal Consciousness to take over. Contemplative prayer can lead to mystical experiences with the Divine Dimension. It is a prayer in stillness with a mystical character, as the prayers take a passive, receptive stance.

*) Thomas Keating (1923-2018) was an American trappiest monk and abbot and one of the co-founders of Centering Prayer
*) Cynthia Bourgeault is a contemporary mystic and priestess

Sitting in Healing Current
A minimum of 40 minutes

This is a receptive form of meditation, to start with we silently formulate our questions, wishes and problems; for the rest of the time we relax into being completely receptive and open ourselves to receive answers or healing.
We can directly receive Divine healing energy in the sense of the primordial energy of the Divine Field or Universe, knowing that this energy comes through spirit beings.

The Healing Current

Usually current meditations are organised by a center for spiritual healing. The action of spirit beings is essential for the intensity of the conveyed divine energy. An egregor pheno-menon can sometimes be part of it, although generally this does not add anything essential to the energy. You can also do a meditation in current individually and request help from a being of the divine field. This form of meditation builds on your trust and belief in a universal intelligence.

Earth-Healing

3.6.20
The elemental of the 5th
kind behind our house left,
as announced a few
weeks earlier, and his
place was taken by a
younger elemental.
He had first appeared
around the time the
Dagda came and linked
to the quartz crystal in our
meditation room.
He had helped me to
establish several hundred
contacts with nature spirits
as well as with beings of
the angelic hierarchies. [1]
These meetings helped
me to a better
understanding of the role
of spirit beings in the
building and maintaining
of nature.

In our endeavors to contribute to earth healing, it may be useful to distinguish between two main forms:

1) **Global Earth healing** (with or without a crystal) and
2) **local nature healing** or neighborhood healing.

Earth-Healing with the quartz crystal

For a number of years, I have been observing different kinds of energy lines appear around a quartz crystal on the carpet in our meditation room. These energy phenomena are perceiveable, with some practice, by anyone present. You may need to develop some way of sensing energy first. The energy structures converging towards the crystal are a perceivable bridge to the invisible and therefore valuable. The crystal functions as a type of communication switchboard that one had in the olden days (see photo). We don't necessarily need a crystal as the lines can also be felt around us. The crystal may make perception easier as the energy lines can become visible in front of us on the floor. These crystals are excellent tools for earth healing, as they create communication lines between us and nature spirits or other type of spirit beings. The Dagda has been helping me in this task. I describe him on my website and in my German and French books on earth healing. [1]

Here is an overview of the different energy phenomena I have observed around the crystal.

Energy signatures around the crystal [1]   (photo)

2 cm thick line	= devas of parcels, landscape angels
30° segment of circle	= large elementals
120° segment " "	= angels of nations
90° segment " "	= angels of ethnic nations
180° line	= global connecting with all very large elementals of a kind

A widening line reaching complete circle = total neighborhood

With the help of my Hartmann-Antennae and my Yes/No questions I am told each time what the being is expecting me to do.

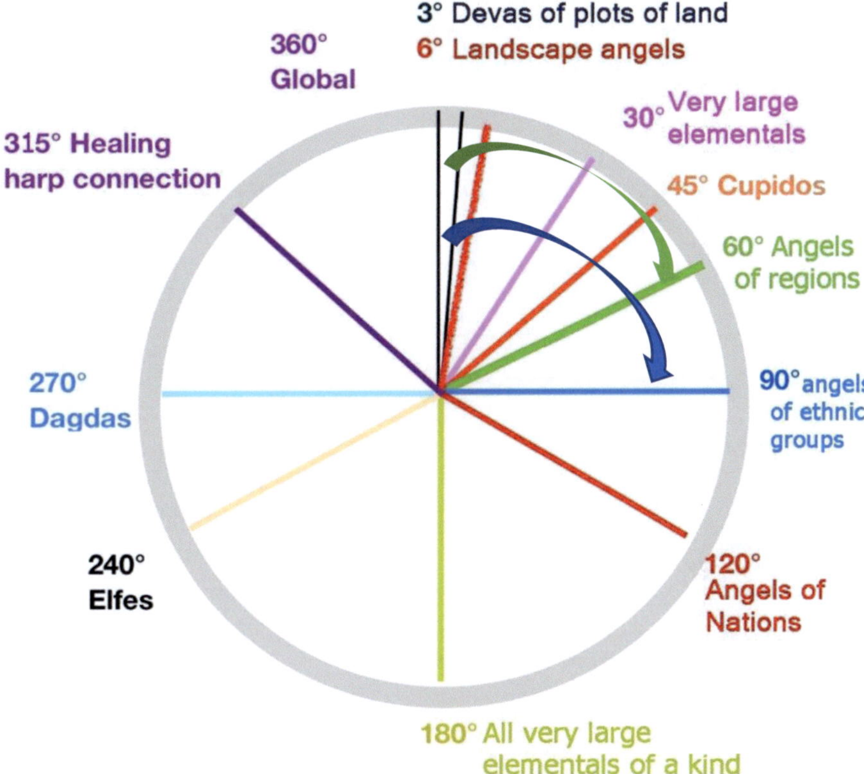

I try to understand what my role is in this healing ritual but much still remains mysterious. I have come to the conclusion that there is a vast intelligence behind all this. I compare myself to an old-time telephone switchboard, where I simply function as an operator who connects the caller and the specialist through a simple plug and wire. The most important part of the connection happens between the sender and the receiver.

Healing or prayer is based on the energy of love and light. Humans have a free will and can decide how and when they want to integrate and apply this energy. It is about growth, a soul's growth. We may wonder why plants, nature spirits and even angels would ask for healing, as they do not have a free will? The many requests that I have received over the years made me think that beings of the natural world need to become embedded in universal love and for example connected to 'coaches' in order to overcome disharmonies that happen through energy changes and evolution. Our actions as physical, feeling and thinking human beings seem to contribute to this balance. In recent years this may be even more so, because we are asked to contribute more actively to and better understand how nature and harmony work.

The following description of global crystal healing is a simplified form of distant healing for the earth. I have described on my website and in my book on Earth-Healing how it developed [1]. I had gone through a long phase where I had detailed contact to many spirit beings who needed my services. Devas, elementals of all sizes and kinds up to regional angels and angels of nations came with requests. This was a fascinating adventure where I learnt a lot about their tasks in maintaining the functioning of nature.

Global crystal healing
10-15 minutes

This type of distant healing is directed towards all beings, that have contacted me on a particular day through the crystal to ask for help.
I also include all other beings and places that have come to my mind in recent time for example through the news.

As with all distant healing I use a short prayer at the beginning. Everyone can choose their own introduction.

'From the depth of my heart and with the help of the Divine Dimension, I pray that the following beings may receive distant healing.'

I am convinced that we only attract the connections to those beings that we are meant to help. In order to simplify the multiple and complex connections to a great number of beings, I came to the conclusion that a global distant healing procedure could include all beings and places I wished and that this would work perfectly.

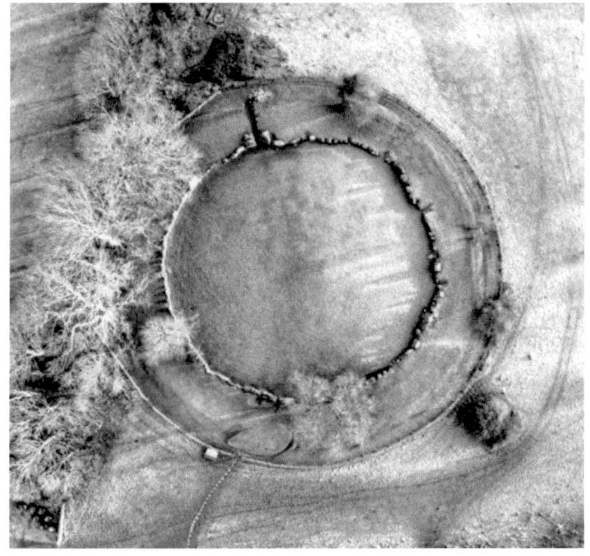

Local nature healing

In May 2020 I was introduced to the new kind of nature spirits that I mentioned earlier - the Lenos-beings. Promoting local nature healing is their task. This local nature healing is different from two similar forms of distant healing: **global Earth healing** and taking care of **our 'private garden'**. In our 'private garden' we know every corner, and can care for every plant – even if this 'private garden' only consists of our apartment, its plants, and flowers. With global Earth healing we cannot do this. This healing addresses all the problem areas in the world that we know of. With **local nature healing** we direct our thoughts to any area or place that we want to focus on . We know our neighborhood and can scan it and visualize it in our memory. It may include our favorite walks along riverbanks, forests, parks, and hills.

We can do this local healing alone or with a group of people, and it is best done for 5-10 minutes daily. It can also include specific distant places such as our favorite holiday place. We must not neglect nature in our local area. Collectively we need to develop a more conscious and compassionate connection to our local areas and nature. This concerns all sentient beings within that area, visible or invisible: animals, plants, trees, bees, humans, nature spirits, etc. This leads to a new, more conscious caretaking of the area we live in.

With distant healing it is not necessary to completely understand how it works, who is receiving it, what they do or what they exactly need. We are simply transmitters, serving other beings.

Local nature healing
Meditation of 5-10 minutes daily

This type of distant healing is directed towards a geography-cal area well known to us and includes all its sentient beings, visible or invisible: animals, insects, nature spirits of all kind from Devas, elementals from the smallest gnome, undine, salamander or sylphs up to landscape angels that take care of a circle of 1 km diameter. It also includes all other invisible beings that work on the good functioning of nature like Elves, Dagdas, magical and mystical beings.

Again, a brief prayer may be used as introduction:
'From the depth of my heart and with the help of the Divine Dimension I pray that the following beings may get distant healing'.

New Nature Spirits - The Lenos-beings

The elemental being situated behind our house had information to communicate today 5.5.20. This kind of contact usually manifests through an energy line leading from his place near our house to the quartz crystal on the floor in the meditation room: 'At the end of May a new type of nature spirits will appear on earth. Their task will be to encourage distant healing from people towards nature. This can be done through 10-minute meditations, during which time you send distant healing to nature, including animals, insects, nature spirits, plants, trees, etc. There will be approximately three of these beings per hectare, which means many new beings. They are being created by the Earth Mother and will remain under her guidance. They will have their own hierarchical organisation led by the Celtic goddess Bríd. This information should be spread. The contact to those Lenos beings will build up over time.'

These new beings were introduced to spirit beings during a series of gatherings that have taken place at sacred places all over the world. [1] Amongst the other subjects discussed during those gatherings were the changes in pollution during the wave of corona virus quarantine, beginning in early 2020. The next gathering will take place at the Autumn Equinox.

7.5.2020
The name of these new nature spirits will be 'Lenos'.

13.5.
Nature distant healing operates mainly through the upper reflector ether by the means of light. The spirit being behind this particular type of distant healing is Saint Bridget or Bríd, the Irish-Celtic goddess of dawn and of the incoming light. Goddesses from ancient Greece and Rome who are

comparable to Bridget are known for the same aspect of Goddess of Dawn. In the Irish-Celtic mythology Bríd is the daughter of the Dagda, the harp player and, in recent years, the creator of the little people, of the small Dagdas, that are connected to the quartz crystal and the wisdom of the Earth.

16.5.
These Lenos beings are guided by Saint Bríd, the 'high one', 'the noble one'. She is the Goddess of dawn, of Spring (Imbolc, light mass), of healing and a daughter of the mythical Tuatha dé Danann. The Name 'Bríd' has got roots in the proto-Indo-European language linking it to 'Berg' (the high and noble one). 'Berg' means also 'Mountain' and is where we can symbolically see far (distant healing); she is the first one to see the morning sun. Healing being traditionally part of the tasks of Saint Bríd.

22.5.
The Lenos beings have a new task concerning distant healing of nature. They will work together with the elementals of the 5th kind encouraging people to co-operate with nature spirits. These elementals should explain to humans the work of the new Lenos beings. I asked the Lenos beings how they plan to go about their work. They will be sending impulses of compassion and confidence to humans and will explain to nature beings, animals, and insects that the distant healing energies they notice are coming from humans. They also will help people to understand how this kind of distant healing operates.

Beginning of May 2020
After sending distant healing
to the beings in our valley
I glimpsed the following
image during meditation

'Someone kneeling
in the high grass
lifted a shallow bowl
up out of the grass.'

A bowl is a female symbol,
that represents receiving
and expresses thankfulness.
I felt the nature spirits of the
valley wanted to show me
that they had been
perceiving this new way of
local nature healing.
I felt that they knew that I
was in the process of
writing about it.
Knowing and feeling that
distant healing for nature is
real and is having effects is
invaluable.

31.5.20
They manifest again, through six energy lines converging towards the crystal. At the moment there are six Lenos beings around our house. Today they tell me that the distant healing I am sending to nature in the neighborhood is having a strengthening effect on the life ether.

7.6.20
As so often, positive beings also attract some opposing beings. I understand them as being obstacles, here to remind us, that in order to get a healing effect we need a conscious effort. A fundamental change in our thinking and being is needed. This is not a superficial formality.

28.6.
They suggest we do local nature healing every day.

3.7.20
A Deva from a nearby forest area has been manifesting itself with a line coming from the South towards the crystal. She informs me that my local nature healing has a significant effect on the light ether of the trees and plants in her part of the forest. (The light ether being the architect of a plant, creating lines of light along which the veins of leaves and twigs then grow their physical substance.)

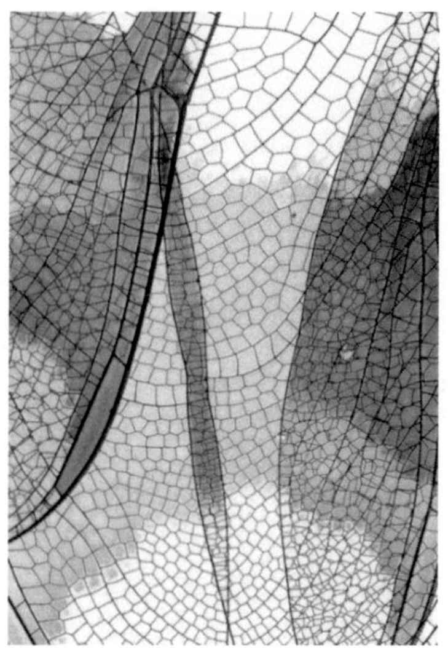

Dwelling in the energy of the Black Virgin, the Earth Goddess

During this meditation I let
my awareness sink into the
lower half of my body,
sometimes far beneath my
feet,
down into the natural
simplicity, into the receptive,
patient welcoming Earth,
the female Creatress,
the resting Blackness Space
of the Earth Goddess of all
times, Notre-Dame-Sous-
Terre
as she is called in Chartres. It
is essentially a space of no
thought, the releasing of all
concepts and preconceived
ideas, the space where real
creativity begins.

Accepting this our
foundation, the basis – body,
earth, grounding – is the
starting point of all progress
and creativity.

Transformation Exercises

The aim of personal transformation exercises in the context of distant healing is the opening of the thyroid chakra (expression) and the heart chakra (compassion). These two chakras are essential for distant healing. This can be done with energy awareness exercises where our higher Self and the upper mental part of our aura structure work into our energy fields in and around our bodies.

These type of exercises draw on the understanding that the physical body is being built and restored according to the blueprints of its etheric. Limiting thought and emotional structures and habits are imprinted into our etheric (our subconscious memory). They can be reprogrammed with the help of our thoughts, our higher self and the beings of the divine field.

The final transformation of the ego is the aim. These kinds of exercises must be designed by a teacher who understands how to bypass and durably transform the defence mechanisms of the ego. In a first phase of transformation a good contact needs to be established to the lower three chakras including a series of points in the lower half of the body. Ego independent geometrical figures, like the isosceles triangle, that acts as as a stabilising form, can be helpful. In my experience in order to obtain lasting transformative results a number of complementary methods have to be used simultaneously (e.g. meditation, devotional practices, spontaneous creative expression, transformation exercises). It is helpful to have a mentor/guide to accompany this. Page 83ff I have described the role of the ego as being necessary as long as the person has not found their individuality. Once this goal has been achieved the temporary structure of the ego can gradually be dissolved.

Spiritual Healing

Spiritual healing work is accomplished with the help of spirit beings, whether we are focusing on the body or in the aura. Spiritual energies can come from specific, known spirit beings or come from the divine energy. In all forms of healing and therapy spirit beings are always contributing whether or not the persons present are conscious of their presence. When energy is at work a spirit being is always with it.

From my book 'Science of Spiritual healing':
"Although it is often an illness that brings us to look for a spiritual, or a deeper healing, the main aim of spiritual healing is to lead the patient back to a state of peace. When a person finds peace within themselves and with the Divine, the body and the illness become somehow secondary. The primary aim is to guide a person into a transformation process and if physical problems can be healed at the same time, all the better. This does not mean neglecting the body or the physical pain but there may, for example, be karmic reasons, that make it is not possible to save or completely relieve the physical for the time being. The progression of the soul, seems always to have priority. "

In his excellent book 'The Art of Spiritual Healing' Joel Goldsmith underlines the importance of relating exclusively to the divine source in oneself and the other. The concentration on the problem level of the receiver should be avoided.

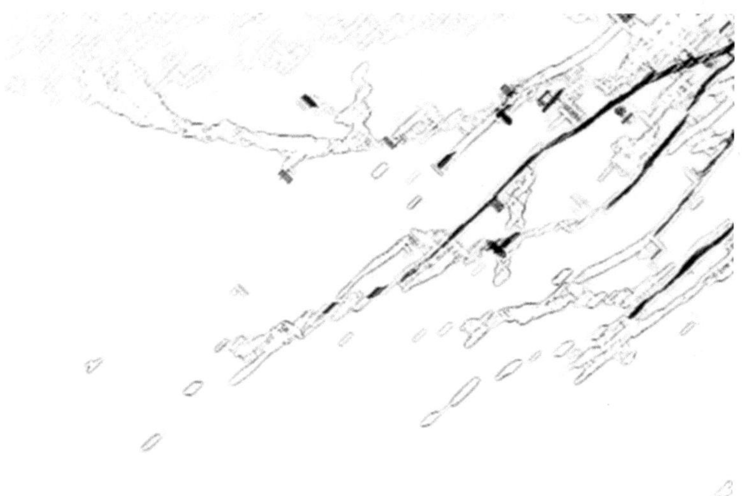

Purity, light, incommensurable
expansion, with no beginning no
end, uninterrupted creativity
construction-destruction-
transformation

are the expression
of the nature
of the timeless
Universal Mind

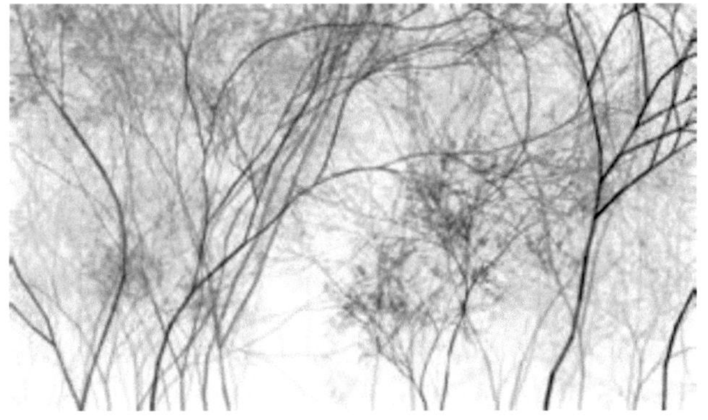

Healing - seen from the perspective of the Healer

Every healing is a meditation, an exercise in centering, observing and being receptive for the healer. Spiritual healing requires us to be open. We can then be guided with the help of our intuition, feelings, and non-physical beings.

We may find ourselves without any known points of reference and often without any ideas of how to continue after the initial steps. Our ability to work without ambition, having quietened our mind, dealing with our own fears and insecurities, all of this is invariably happening to us during a treatment.

Bob: "The main thing is that the person who receives healing, is contributing something. Without this the whole thing becomes a pure act of will that, in my opinion, has got nothing to do with spiritual healing, but a lot with an activity of the etheric body."

Radiesthesia and Radionics

Depending on the country and the tradition there are different definitions of radiesthesia and radionics. A simple definition is that every person is a receiver /radiesthesia practitioner and a sender / radionic. The technique of radiesthesia can be used for diagnostic purposes. With radionics energy fields can be changed. Both radiesthesia and radionics also work at a distance.

Radiesthesia: According to 'C' the perception of energy and information, mainly happens over the upper mental area of the aura and the intuition of the third eye. In Radiesthesia treatments practitioners often use also lists and diagrams. In German speaking countries Radiesthesia means a method of perception. The word Radiesthesia means being sensitive to radiations. The perception can happen with or without the help of instruments. The instruments commonly used are the pendulum, the rod, or H3-Antennae, which is a technical tool used in frequency centered radiesthesia. Any perception yet always has an influence on what is observed.

The word **radionics** is used for the sending of energies and information. These can be healing energies, thoughts, patterns, and frequencies, that are sent to assist people, animals, or plants. It functions with the help of the upper reflector ether layer, that operates on a planetary level as a kind of transportation pipe. In this form of healing, spiritual or divine energies as we have described with prayer and distant healing, are not necessarily in the foreground. The reason that may be that classical radionics tends to use mental structures such as symbols, lists, schemes, sometimes technical machinery, which can lead the practitioner to rely too much on intellect. Also, in this form of treatment spirit beings are always co-operating. The inner motivation of the healing practitioner attracts corresponding spirit beings and determines the different components of our 'sending'.

These subtle energy techniques may give the practitioner the impression of being able to perceive and heal all the presented problems of the client. Any healing treatment always needs to respect the natural process of development of the receiver. It is a delicate question to know when an intervention is appropriate. We may ask ourselves: are we able to do the intervention? Is the person agreeing? Is it in the deeper interest of the person's soul and therefore is the spiritual dimension agreeing as well?

***

This quote from my teacher contains so much, that, after all these years, I am still discovering more details of it.
Bob Moore: "The reflector or warmth ether contains the feeling between people, it contains world memory, knowledge, wisdom, including physical memory, intuitions. Thoughts enter here before entering the brain. The reflector ether is the highest ether, the most advanced contact we can reach with energy. In this contact with energy you have all the things that are necessary: peace, contentment with yourself, you have the connection with the universe ... and finally you have the balance between nature and yourself. The reflector ether is the combination of all of this, but essentially *centered around peace*."

Bob Moore (Spiritual healer and teacher; 1928-2008)

This upper layer of the reflector ether has, on a planetary level, a layer of energy along which healing energy travels. It is also, in the human etheric, an interface between high energies like love, peace and physical body and our mind.

Healing of places

Whilst writing a book on 'churches of all times' [8] I visited many buildings and pre-Christian sacred places in order to understand their energy. To my astonishment I discovered that a considerable number were energetically in an unbalanced state. The thousand-year-old history of church buildings in Europe has often left traces of events that have not been healed. In most cases the local community has to participate in the healing. Only in one case, where a crime had happened in the isolated chapel in a forest 800 years ago with no links to the local community, could our group successfully harmonize and heal the place, using prayer, singing, and by asking for forgiveness.

I recently participated in the healing of an island that had been cursed and was situated in the middle of a town. In the early 13th century the island had been used as an emergency camp for people infected with leprosy or plague because the nearby hospital was full. It would seem that medieval beliefs led the town people to curse the island and the river. This had happened because the waters had been contaminated by the improper disposal of waste and sewage which had caused further contamination and had increased people's fears. A fallen angel had established itself there when a type of exorcism ritual had caused the death of a young woman. Her soul had become lost in eternal revengeful actions (astral healing) ever since. The healing involved asking for forgiveness for all persons involved, in order to lift the curse. It also involved freeing the young woman's soul and soothing an upset water elemental who had been wrongly accused. The healing procedure was done at distance and with the help of some local healers who had a personal connection to that island.

3. Reflections

The effects of distant healing and prayer

Distant healing and prayer are above all a matter of trust and belief. Some people continue to see the half empty glass, others see the glass half full. Looking for a generally accepted proof of the value of distant healing and prayer is like wanting to prove the existence of God or the efficiency of homeopathy. The perception of subtle energies and its interpretation is helpful.

Our beliefs and thought structures determine our experiences. We each have our own free will to walk on the paths we choose and then to experience the consequences of our choices.

My perception of subtle energies has brought me unequivocal proofs of the effects of prayer and distant healing. In my book 'Earth-Healing' and on my website I have described in detail how communication lines coming from nature spirits towards my crystal have disappeared once I had sent distant healing. These effects always linked to a direct communication with the nature spirits. In my booklet 'The Harp in Distant Healing' I describe how the effects of distant healing with the harp can be observed, creating a widening circle in the reflector ether as perceived on Google maps. These observations can be confirmed by persons present in the room.

My co-operation over many years with spirit beings has brought me the continuous proof of their existence and of the existence of subtle energy.

How does healing and prayer work at distance

As soon as we widen our own horizon and begin to feel the suffering of sentient beings in our hearts, we wish to contribute to helping them. Deep inside we know that we are all in the 'same boat', feeling that the suffering of someone else is also our suffering.

Quantum physics explains that we are all connected through the tiny particles within all our atoms, called Quantum entanglement. A pair of entangled particles once connected remains strongly correlated over any distance. Any change produced in one atom instantaneously creates a change in the other atom and hence defies the theory of relativity, according to which no effect can propagate faster than light. (source: Unfoldanswers.com)

The mental energy of the third eye programs the address we are aiming for (name, photo, geographical spot), the energy of love is contributed with the help of the Cherubim angels, the divine quality of the Universal Consciousness. Any effect on humans, animals, and nature spirits, at a distance is happening with the help of the reflector or warmth ether (see page 105). The upper astral quality of compassion plays a role in this as well with the all-encompassing, motherly energy of the Earth Goddess/ Black Madonna. The whole process is accompanied by spirit beings. It helps to ask them for assistance.

Even after this brief explanation from C the mystery remains. I believe that distant healing works in an intelligent and complex way, even if I do not completely understand it.

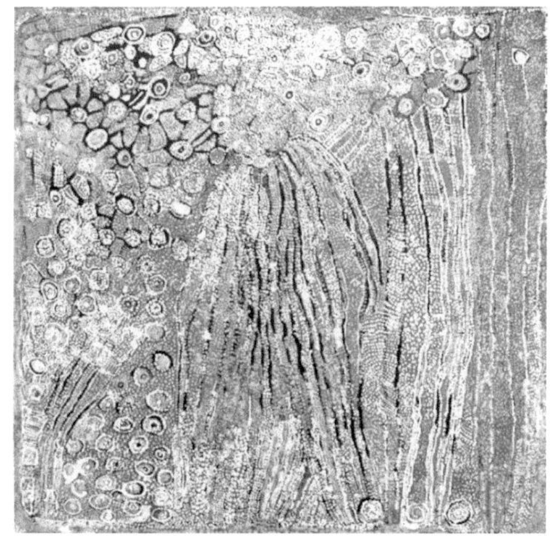

Definitions (second part)

I will quote extensively from the Christ letters over the next few pages as they bring clarity into what healing is about and how and why disharmonies have happened. They clarify the foundation upon which this book is based.

"I was aware of the Divine Harmony controlling the work of the cells, which were busily building and maintaining the various parts of physical bodies of all living creatures and plant life, big and small. This is why I drew heavily on the countryside to give examples of the immanence and activity of the 'Father' within the least of wild life – such as plants and birds."
Christ letter 5 / p. 6

"I saw that the true nature of 'God Mind' was the very highest form of Divine Love and this could be seen consistently active within every living thing." p. 7

"In music, in the sea, in a flower, in a leaf, in a friendly gesture, I see what people call God – in all these things." Pablo Casals

"…the 'Universal' and 'Divine"… "I also 'saw' that creation was a visible manifestation of the Universal Creative Impulses of Being…" Letter 5 / p. 1

The Christ letters were channeled in 2000/2001 by a then 80-year-old English woman. The channelings are in the original English version 97% pure, according to 'C'. This is unusually rare. The German translation e.g. is not everywhere perfect but still reaches 95% perfection. Independently of the name of the channeled being, Christ according to 'C', the texts are of extraordinary quality. [5]

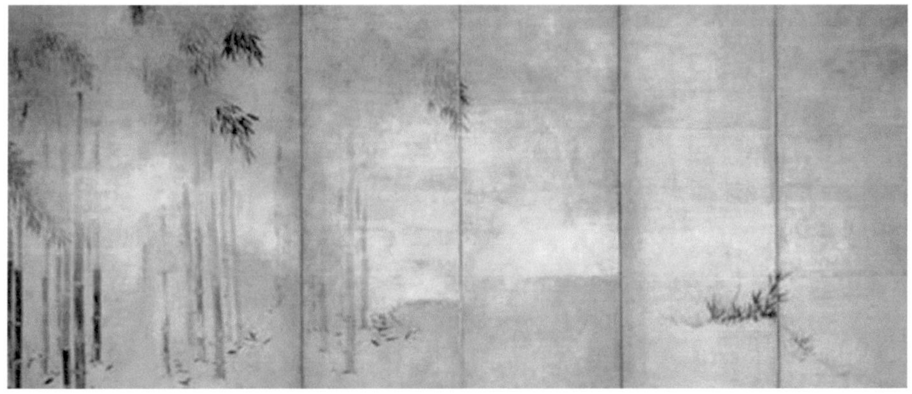

Spirit/Mind

Conscious thinking and feeling controlled by our higher Self and individual will.

The word 'spirit'(German 'Geist') is very close to the word 'mind' and includes a multitude of meanings: opinion, aim, intellect/thought, meaning (ability to think and reason, ways of thinking), soul, fantasy, memory, focus of thought, attention, mental sanity, the mind of the universe, etc. It refers to a higher, invisible, and intelligent directive force. The term 'higher Self' may include the co-operation with and impulses from beings of the divine field.

From my book *Evolution of a Soul* (in German) [6]: "from Greek *pneuma, nous, psyche*, Latin *spiritus,* French *esprit.* The terms 'mind' and 'spirit' have been used for historical reasons in non-uniformist ways in philosophy, theology, psychology and in daily language. The confusing diversity of use of the term 'spirit/mind' is due to the fact that we have great trouble to grasp the invisible and to distinguish between what we perceive within that invisible field."

Consciousness

Capacity to experience oneself as a perceiving, feeling, independent being, with a will and a purpose. Consciousness is the sum and the content of all our perceptions.

This quotation comes from my book *Evolution of a Soul* [6]: „Awareness is the conscious and focused part of conscious-ness that we direct to an object. Consciousness includes more than just a thought consciousness, as it contains also feelings and finer perceptions."

David Chalmers (TED talk 2014) suggests that consciousness is a fundamental property of the universe, like space, time, mass, and electromagnetic charge. Each system, each feeling being has a degree of consciousness of its own.

This leads me to the hypothesis that also all groups, is each a system, have their own group consciousness. The planet earth is also a system, as is an entire river system and a mountain and thus also has a group consciousness. This way of thinking determines our connection to nature, to animals, insects, all sentient beings including invisible nature spirits of spirit beings of the divine field.

Whole – Holy

Wiktionary: from old Norsk *heill* (färör *heilur*, Norwegian, Danish, Swedish *hel*, proto German *hailaz*; English *whole, hale,* Dutch *heel,* German *heil),* from proto-Indo-European *kohilus* (*healthy, whole);* used also in 'healing waters', 'healing herbs'.

The All-Oneness and the reason for losing this connection can be understood when reading the following passages in the Christ letters. The word holy is connected to the concept of Whole. Holy refers to a place or person, that knows and encourages the connection to the Whole, the universal creative force.

Free will

We can only really respect a person's free will when we have some understanding of the purpose of an earthly incarnation. I have met radionics and radiesthesia practitioners, therapists and healers who thought they had to fulfill all of the person's requests. This can easily be an illusion as healing is happening from a spiritual/soul level. We are all simply servants of the Divine.

Rudolf Steiner said that in our time we are developing two inseparable values: Love and Freedom.

Although as a healer we contribute to a person's growth, we must not take away their own responsibility for working on themselves. I believe that ultimately everyone is responsible for what happens to them. In the case collective disasters such as famines, wars, and environmental problems this is not easy to understand. This does not mean that we cannot assist someone, by sending them healing or praying for them.

Divine Harmony

…or the state of 'Complete Intelligent Love'. If in healing we want to grasp what Ultimate Universal Harmony is, we need to get some understanding of what has separated us from that original paradisal state. This separation was not due to 'sin' in the sense of becoming guilty. It seems to have been an inevitable step in the plan necessary for human evolution to happen. In the following thoughts we will go beyond the conventional psychology of childhood events, even depth psychology and reach a spiritual dimension of psychology in order to understand the deeper mechanisms of evolution. The Christ letters express it masterfully.

"Man's inbuilt impulse to protect his own individuality had made him set up rules and laws for human society. The 'Universal Creative Power' – LOVE – had absolutely nothing to do with the setting up of human restrictions, limitations, laws, and judgment. …

"Only mankind understood the meaning of the word 'sin' since only mankind and all of 'creation subject to mankind', 'could ever know the pain, deprivation and misery caused by the two fundamental impulses of Individuality – **Bonding-Rejection** active within the human 'personality'. "

"The entire universe is a manifestation of the 'Creative Power' active within these twin impulses of physical being – creating 'matter' and individual form. This is one of the fundamental 'secrets' of the universe." Christ letter 5 / p 25

He explains further the two basic impulses of Binding-Rejection as being two faces of the same coin:

"The face of BONDING drags, draws, attracts, demands, pulls, buys, grabs, clutches clings to the people and possessions it craves. This impulse creates an illusion of security in togetherness and possessions.

It is the 'tool' of 'Mother Consciousness' inspiring the building of families, communities, and nations. It can be productive of beauty, joy, harmony, and love. It can also wreck lives and destroy communities when it is 'ego' driven.

The face of REJECTION repels, thrusts aside, pushes away, evades, everything – people, animals, possessions it does not want. Thus, the impulse of rejection creates an illusion of privacy and security.

It is the impulse that urges rifts in families, relationships, communities, and nations. It is supposedly geared to saving lives, ensuring protection and privacy but is a destructive force when it is 'ego' driven.

Without these twin impulses of being, all things would have remained forever merged into one another within the eternal timelessness of 'universal creative power in equilibrium'. Without these twin impulses, there would be no interplay of 'give and take' and 'pull and push' necessary to the creation of the millions of personal experiences out of which 'personality' grows and evolves. Therefore, the problem of 'personality' and the 'ego drive' endured by all living things and mankind was/is an irrevocable, unavoidable fact of creation. Any other explanation is pure myth.

The 'Father-Mother-Creative Power' – LIFE – continually flowed through all the universe and was the life using the twin impulses of thought and feeling. Hence any powerful 'imperfect thinking and feeling' could disturb and change the 'consciousness pattern' of created things.

Inversely: My thinking when fully cleansed of the twin impulses of 'ego' – and fully receptive of the 'Father-Mother Creative Power' Intelligence/Love would re-introduce the conditions of 'perfect intelligent love'. Letter No 5 / p 27

Therefore, a condition previously made imperfect as a result of 'imperfect thought' could be brought back into a condition of 'wholeness' again by changing ego attitudes and thoughts to those of unconditional love. My mind was a 'tool' of the whole creative process originating in the UNIVERSAL.

I saw that what men called 'sin' was the direct result of the inter-play of the Bonding-Rejection impulses within human nature." Christ letter 5 / p 26 [5]

Collective forms of energy

Egregors

Prayer and distant healing are activities of our mind, that is our upper mental (intuition and other higher mental faculties). I have mentioned some positive collective thought forms on page 10 in connection with collective actions. When these become charged with emotions, an egregore is created. This part of the collective energy does not generally bring lasting healing effects. These are charged thought forms. The purest form of distant healing and prayer happen when we connect to the divine within us and these energies connect with the divine in the receiver. In this case we avoid the creation of egregores.

We find egregores everywhere in the memory of places, that is in the life ether, also known as Akasha Chronicles. In pilgrimage places such as Lourdes in France this phenomenon is particularly evident. Statues of saints, icon paintings, photographs of film stars, music and sport celebrities accumulate thoughtforms with an emotional charge around them. You can easily observe this in churches where for example popular statues accumulate this sort of energy around them. If a church or similar place is used mainly for funerals and as a 'wailing wall', the negative or heavy egregore aspect would dominate. On the contrary when the place is used mainly for weddings, baptisms, and concerts the positive egregore energies will be predominant. The back section of churches is often less charged as the activities usually take place closer to the altar.

The determining factor in distant healing is the motivation of the sender. We need self-awareness and sincerity in order to be aware of our motivation. In order to pray and to send distant healing we need to overcome our subconscious limitations.

Sending distant healing with a group is sending energy through the upper reflector ether directly to the receivers for example when their names are on a distant healing list. If the group distant healing is connected with a symbol, in the case of a healing center with a leader or saint, a part of the energy will go to an egregore. In the best of cases the uplifting, positive and the negative components of that egregore will be in balance. By 'negative' I mean being overcharged with people's projections of pain, despair, and hopelessness. I use the word 'positive' to mean the connection to upper astral feelings dedicated to the leader, the saint or symbol. This is a complex subject because the admiration of a master, saint or even a godlike figure can contain aspects of self-rejection and rejection of our own divine nature, contributing thus to a negative egregore. If the building of an egregore cannot be avoided, it would be good to be aware of balancing the two parts.

Distant healing, prayers or meditations can trigger similar phenomena, when done in large groups, as social medias nowadays make possible. One can prevent emotionally charged egregore energies by clearly explaining the 'spirit' of the planned event beforehand.

Effects on the sender and praying person

When we connect positive, uplifting thoughts to others this also contributes to our own wellbeing, as we are in contact with spiritual energies. All wisdom traditions have known this: the first steps to improving our own well being is to refrain from all criticism towards others and to do what we can to help.

The perception of other types of effects is different for each person. As so often we cannot compare ourselves with others. Personally I do not experience much during distant healing; especially when I send distant healing with the harp. This has made me look carefully for the effects that I have described in this booklet. Writing this book is the result of practising distant healing for many years.

The feedback I have received from others about the distant healing I send has inspired me to research how it actually works. I teach spiritual healing and am writing about my work and experiences to make this available to others. One of the big questions that I have about spiritual healing, distant healing and prayer is: what would have happened to me healthwise, if I had not received healing ? This leads me to a fundamental aspect: when we seek to strengthen our trust and faith in the spiritual dimension we often have to move through and beyond doubts. Trust and faith are goals or spiritual qualities that do not necessarily bring tangible proofs but rather convictions and can build up through our own life experiences.

A friend shares her experiences of **distant healing**:

"I have been sending distant healing for nearly forty years. At the beginning of a distant healing period I use a short ritual and pray to all the four directions, then pray to what is above and to what is below. This gives me a feeling of space and peace that expands in all directions.

I often start by saying prayers out loud, prayers for the earth and for all beings. I pray for disaster areas, for the security of nuclear power stations. I often feel love energy and light in my whole being. I feel connected to the non-physical world and to people everywhere on earth who are praying for peace and harmony.

A friend asked me recently to send healing for her physical problems and during distant healing I sensed that she was surrounded by light. I tuned into a young woman in the days leading up to her giving birth and felt so connected to her that I experienced many feelings in my own body. When the baby was born, I felt an immense love and joy that stayed for several days. I also pray for people who have died, for about six weeks after their death.

When I send distant healing, I often feel as if I am sinking very deep into the earth and experience a joyful expansion in the heart area and feel that love and compassion are radiating outwards.

I have often felt connected to other people who were sending distant healing at the same time. I can also feel when I receive distant healing myself and that is a feeling of great peace."

Other forms of distant healing

Everyday oportunities

When we see people in the street we can silently wish them a good day. We can wish that during the day that they may see the beauty of nature, of a flower or the smile of a child. Our thoughts send out a powerful energy wave, that also influences us in return.

Positive energy can be transmitted during a telephone talk with a therapist or a healer, even during a talk in the same room. Again there is an action onto the energy fields of the other.

Auragrams and similar forms

An Auragram is a drawing of the human aura. Working at distance in the aura of a person through using symbols is also a form of distant healing. In my understanding this happens through the use of intuition and clairvoyance that links to the upper mental, compassion that links to the upper astral and the third eye chakra. Usually there are spirit beings participating in any healing. The quality of the attracted spirit beings corresponds to the level of consciousness of the healer. Our motivation when sending healing through an aurogram is determined by our thoughts. Are we in tune with the divine energy of Light, Love and Truth? The priority must always be respecting the free will of the person receiving healing.

When I work in the energy field of a person with the auragram structure below, the person sometimes feels what is happening in their aura, whether they are in the same room or not. Points, symbols, and energy lines play a role in this work. I can sense the presence of structures or spirit beings in the energy fields of a person as accumulations of energy, that I detect intuitively , with my hand and my Hartmann-Antennae.

I co-operate with spirit beings, usually C, when working with an auragram. They show me the areas in the aura that the person needs to focus on in an intelligent sequence. These are always areas that the person is able to tune into and work with. We simply point to an open door here and there. The person receiving the information has to move through that door when they are ready. The work of trnasformation is never finished.

Graphism next page:

We are part of all these layers that penetrate us.

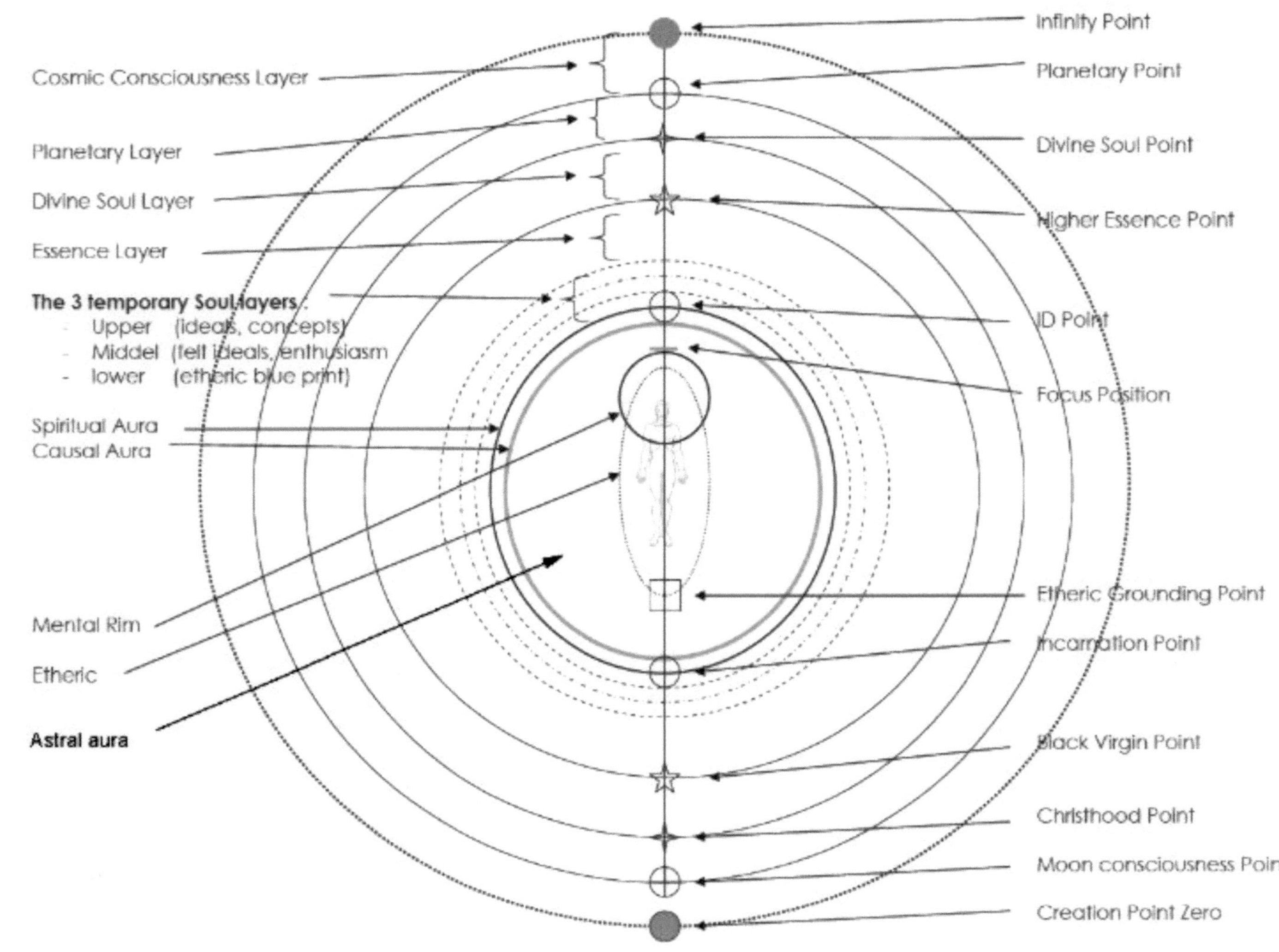

Cosmic Consciousness Layer
Planetary Layer
Divine Soul Layer
Essence Layer
The 3 temporary Soul layers:
- Upper (ideals, concepts)
- Middel (felt ideals, enthusiasm
- lower (etheric blue print)
Spiritual Aura
Causal Aura
Mental Rim
Etheric
Astral aura
Infinity Point
Planetary Point
Divine Soul Point
Higher Essence Point
ID Point
Focus Position
Etheric Grounding Point
Incarnation Point
Black Virgin Point
Christhood Point
Moon consciousness Point
Creation Point Zero
Daniel Perret - Contributing to Healing

Auragram by 'C' of the corona virus situation

On March 21 2020 I asked 'C', the college of spirit beings from Rocamadour, to comment about the corona virus situation. I decided to make an auragram to illustrate their thoughts. In the center I usually place a single human being. This time I asked for the whole of humanity to be put in the center of the auragram. All around are the energy layers of the aura reaching out until the cosmic layer. I asked C to show me the important components of the situation.

The results: C showed some spirit beings, egregores, and places in the collective energy field, here exclusively located on the vertical axes. This is the central axis connecting heaven and earth that goes through the spine and thus through all our chakras. The sequence in 'C's explanations is always meaningful, leading us from the cause to the steps needed for healing. Each of the four causes 1 – 4 is perceivable on the auragram as an accumulation of energy that I can detect with my hand or the Hartmann-Antenna. In a concise manner C brought all the essential aspects to my awareness: human misconceptions, ecology, immune system, empathy, and the illusory tendency to separate ourselves from the whole. For the explanations on the points and layers see [9].

The results:

1. At the point of the divine soul C shows us a collective thought form, an egregore. This is made up of thought structures and emotions that have built up over the last 400 years and are centered around our materialistic Newtonian* view of and connection to nature and the earth. These thought forms bear witness to our lack of respect towards the earth and this has directly caused the virus. *) Isaak Newton, English scientist who laid the foundation for a rigorous materialistic physics.

2. C then points at humanity's root chakra, which is our real connection to the earth, to the ground, nature, our bodies, and to our life circumstances. These all need to be rebalanced.
3. Then C points at humanity's heart chakra. This is the place of lasting transformations towards more empathy, that we need to achieve individually and collectively.
4. Finally, C shows us a challenging being in the etheric blueprint. These spirit beings from the galaxy Redshift 7 want to underline the illusion of us trying to separate ourselves from other people, countries, nature spirits, viruses and thus from the above causes 1-3 using walls, masks, prejudices, artificial separations, and suchlike. These beings have placed themselves in the blueprint of our human etheric. Our etheric energy field holds the first phases of any physical illness. Our erroneous ways of thinking (see egregore at point 1) have caused a weakness in our immune system, first of all in our etheric. This is the result of the causes 1-3 and can only be healed through a change of attitude towards the earth.

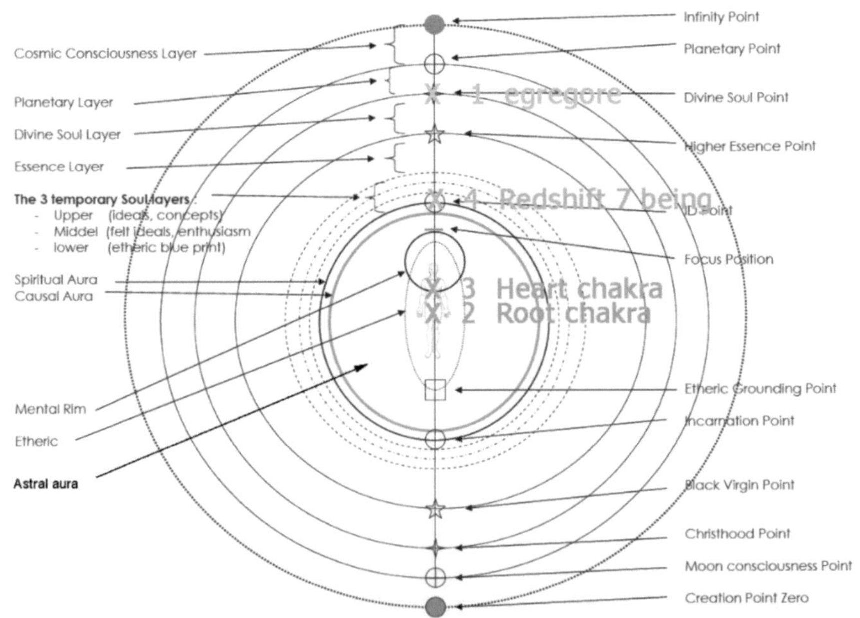

Transpersonal Art

Transpersonal art leads us into levels of consciousness beyond the personal. Many transpersonal artists connect to spiritual dimensions. Their creations can convey hope and spiritual upliftment which bring courage, joy and healing that is often on an unconscious, transpersonal level.

As an improvising musician, I have moved for many years in comparable spaces. Improvisation in a spontaneous, authentic artistic expression is a door into the transpersonal when our transformational work has reached a progressive state.

Album '**Earth Healing**'

'**Magical Squares**' are found in churches and cult places of all times. They are divine seals and carry a considerable amount of energy as well as a great potential for spiritual development. [4]

Danielperret.bandcamp.com

I now include two of Marie Perret's intuitive paintings and accompanying texts. (pages 97 and 101). Both paintings are taken from her new book 'Painting into Life'. [3]

Ascension towards the Light   2015   1m x 70cm

On 28th December 2014 an airplane crashed in bad weather in the Java Sea and 162 people on board were killed. I was saddened by this news and included these people in my daily distant healing. Several days later I went into my atelier, not thinking of this event and with no idea of what I would paint. I began to paint a dark blue background and soon felt as if I was in waves, dark waters and night. The large light figure with outstretched arms and peaceful face emerged and as I continued painting, I was filled with love and began to see the small figures rising up out of the water. I then realized that I was connecting to the air crash. I then sensed many other light beings/angels arriving with open loving arms to receive and comfort the people who had died and guide them up towards light.

Automatic writing: You are never alone - we are watching and ready to come to the aid of those in need. We cannot and do not interfere with the free will of humans but when there is a need and a request for help, we intervene and assist in any way we can. Life does not end with physical death. The release from the physical body that you call death is a transition to another state of consciousness and life continues in other dimensions where much healing and growth are possible. When tragedy/ disaster such as this happens and many people die in shock and fear we assist the souls at this important threshold.

We hold you in love and tenderness, light and softness to release fear and contraction, and allow the passage between worlds. We soothe shock and fear, we work with grieving families and open a pathway of light out of the dark waters.

Our Light, our love is healing the world.'
2020   1m x 70 cm

When we send daily distant healing, I imagine that millions of people all over the world are doing this at the same time and at many other times during the day. People are praying for peace, love and harmony and I imagine this as a network of light shining brightly all around the earth.

I feel that the collective strength of our individual prayers is expanding a softness that is affecting nature and all living things. This softness is penetrating the hearts and minds of all human beings. The vision that I hold in my soul, the hope that I carry in my heart spontaneously emerged in this painting that I call 'Our light, our love is healing the world.'

An increasing number of people have made an inner journey to open their hearts. They are walking and praying, and singing to each other through the landscape. They are standing like trees on the solid ground inside themselves with open arms extending their love outwards. Their heart prayers are scattering seeds of light into the atmosphere that surrounds the earth. These seeds spread out and settle in others' hearts and minds and in time they take root and grow into positive thoughts, words and actions. New shoots are pushing up through cracks in the concrete, growing on wasteland, in areas that have been burned or on land that has flooded.

The central figure is the Divine Feminine. She is gaining strength as people open their hearts – she is within each of us, she is you and I as we heal ourselves, overcome our fears, and embody love and light in our words and actions. So many positive initiatives are happening today, so many inspired people of all ages are working for change and justice…we must visualise and trust that this is happening. People are waking up and opening their hearts to the reality that we are all 'brothers and sisters' who share common needs. We are weaving golden threads of spiritual connectedness that will eventually link up into a spiritual safety net for the world.

I hold the vision that goodness and co-operation is growing and will eventually overflow and tip the balance – peace, love and light will return to our planet.

Postscript

Healing, in the sense of feeling that we are part of the Whole, the All One, starts with ourselves and expands to our immediate surroundings: to people, animals, bees, insects, birds, air, nature, water, and nature spirits. Think global and act local. I invite you to extend your awareness beyond your private patch, beyond your garden, or apartment, and to include your surroundings in your hearts and contribute something positive to it.

Bob Moore's indications that the reflector ether is the highest energy contact we can achieve (page 71), means we can work towards attaining that. Considering that this reflector ether is so instrumental in distant healing and prayer, it shows us the way forward.

The 21 Spheres of the Divine Field

This list has been elaborated together with 'C'. It allows us to get a more detailed insight into the invisible worlds of the divine dimension. At the same time, it reshapes an antiquated idea of God. The beings of spheres 17-20 can be considered as the ruling, directing angels; those of spheres 12-16 as the executing and administrating angels, the angels of level 1-11 as the facilitators and go-between angels bridging the gap between humans, angels and nature spirits. I avoid the word levels and use instead spheres in order not to impose a value but rather to see them as spheres of inspiration.

1   landscape or area angels, Dagdas & Devas
    spirit of time, elementals of the 5th kind
2   individual or group karma; angels of small lakes, rivers &
    deserts (Namibian desert angel)
3   wisdom and experience of nature spirits; Kali (breaking
    down & renewal); angels of larger lakes and rivers;
    strato cirrus cloud angels; regional angels; angels of larger
    deserts (Gobi & Sahara); the very large elementals (of
    earth, water, fire, air); beings of volcanoes; the Tuatha Dé
    Danann Dagda (head of the level 1 Dagdas he had
    created)
4   inspirations from shamanic light beings
    Indonesian tropical forest angels; some non-intrusive ET's
5   Sophia (wisdom); life force; wisdom of animals; angels of
    large tropical forests (Amazonian, equatorial African)
6   nation angels: art inspired by them and which contributes
    to shape the spirit of nation (symbols, flags, hymns, etc.)
7   angel or spirit of a continent: Europe, Middle East,
    North America, Arica, etc.
8   respect for life and creation; angel of science, knowledge
    & education; friendly ET's like those from Pegasus;

underlying structures of life; ocean angels: Atlantic, Pacific, etc.; smaller mountain range: Pyrenees, Appalachians

9   St. Bridget/Bríd - Goddess of Spring and Poetry; larger mountain range angels: Ural, Himalaya, Alps, Andes, Rockies

10 planet Earth Angel

11 church Angels: protecting altars, statues, crucifix, etc., Archangels: Prayer/Sound (Sandalphon); good news & creativity (Gabriel); hope & aspiration (Ramiel); service (Jehudiel); faith & spiritual strength (Saraquiel); justice & harmony (Raguel); wisdom (Raziel); transformation of the shadow (Binael); of healing (Raphael); teaching (Uriel), Light into shadow (Michael); Divine will (Hesediel)

12 Archaï – Lordships, Principalities – leading the earth leaders, people and communities

13 Kyriotetes: instructing the duties of the angels below them; their energy is pure grace; Transmission of the Teachings of Christ, teachings of aliveness, spirits of wisdom; Ignacio de Loyola, St. Francis of Assisi, Theresa of Avila, St. Hildegard, St. Cecilia, etc.

14 Exusiai - Powers: protecting the heavenly spheres from all negative influences of the earthly spheres. Keep the world in balance and especially the balance with the dark forces; Spirits or creators of form

15 Dynameis - Virtues: directing the cycles of the planets; Spirits of house baptisms

16 Dominions: leading the Angels of the earth, of continents and nations

17 Cupides : high Angels of Arts, Love and Beauty

18 Thrones: Angels or spirits of divine will and life energy, giving impulses of direction for humanity

19 Cherubim : Angels or spirits of harmony and wisdom

20 Seraphim : Angels or spirits of light & fire, igniters

21 Black Madonna, Christ, Creator, Holy Spirit, Buddha, etc.

Measuring subtle energies

Conventional physics concentrates on one type of 'waves' and knows very little about the existence of other wave forms. The discussion around scalar waves introduces new types of waves.

"The spirit beings of C speak of more than 50'000 levels in the etheric energy fields alone! We must see that, when it comes to subtle energies, we know very little. The complexity of the etheric levels alone is unfathomable for us. They are a wonder of creation. C explains that the different types of waves are: vertical (conventional physics), horizontal for the spiritual waves, diagonal (bottom left to top right) for the etheric levels, diagonal (bottom right to top left) for the astral level, spiral type for the energy of love, and in small packets of information for the mental energies." [4]

The layers of the outer etheric

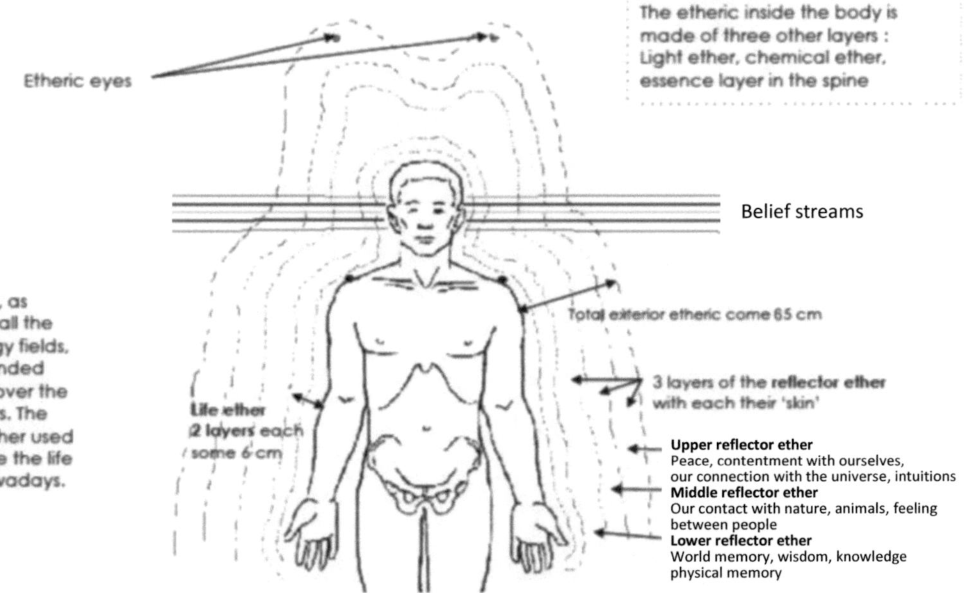

Holistic healing
and the factors of our present well-being

These are general indications of 'C'. Each person, each individual situation, each moment in time is of course different.

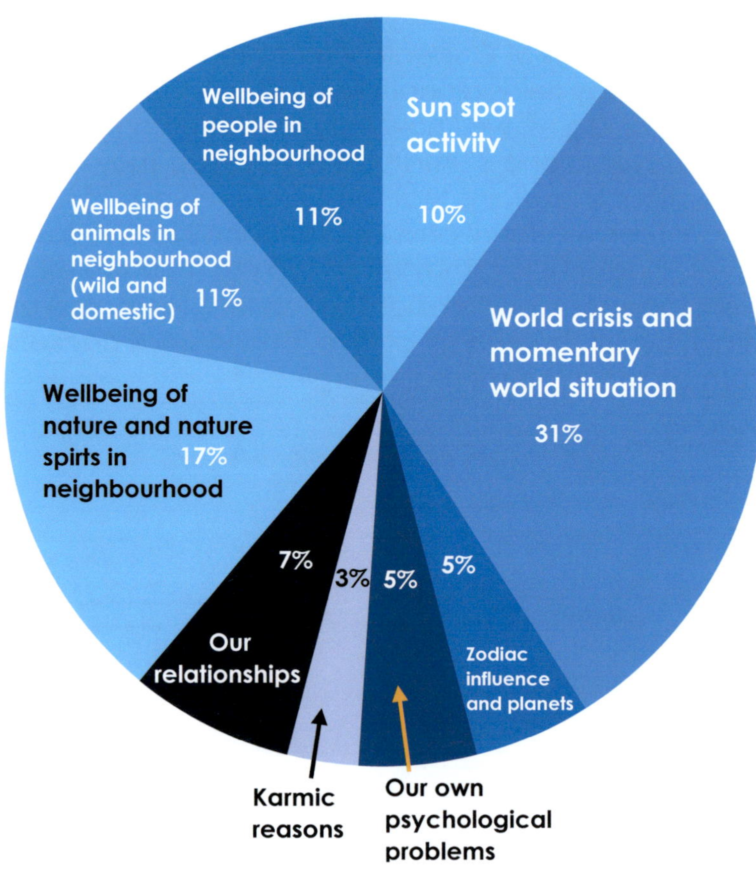

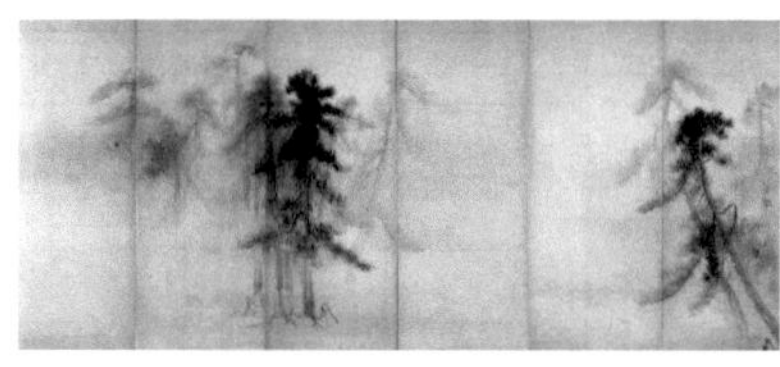

Illustrations: details from three paintings of Japanese painter Hasagawa Tohaku (1539-1610)

These Illustrations are from Maurits Wouters, Studio Airport, Utrecht. Previously published in 'Emergency Magazine'.

Photos D. Perret. Photo of the two birch trees A. Binetruy

page 97, Harp Mulagh Mast Modell 17th or 18th century. Collection National Museum of Ireland. www.rumor.pt

Book List   *also see my Website*

1) Perret Daniel **Erd-Heilen / Guérir la Terre** 2019, BoD
2) Daniel Perret, 'The Harp in Distant Healing', 2020, BoD
3) Marie Perret, 'Painting into Life', 2020, Pumbo
4) Daniel Perret, 'Magische Quadrate – göttliche Siegel', 2020, BoD
5) The Christ Letters, anonymous channeled, christsway.co.za
6) D. Perret, ‚Evolution einer Seele', BoD
7) D. Perret, ‚The Science of Spiritual Healing', BoD
8) D. Perret, ‚Eglises de tous les Temps', 2020, Pumbo
9) D. Perret, 'A wider Self', BoD
10) D. Perret, 'Music as a Mystical Journey', BoD
11) D. Perret, 'Creating Divine Art', BoD

Celebration - Gratitude - Service

www.vallonperret.com